Mammoths

Giants of the Ice Age

for Eleanor & Frankie

www.bunkerhillpublishing.com

First published in 2004 by Bunker Hill Publishing Inc.
26 Adams Street, Charlestown, MA 02129 USA

10 9 8 7 6 5 4 3 2 1

Library of Congress Cataloging in Publication Data available from the publisher's office

ISBN 1 59373 018 7

Designed by Louise Millar

Printed in China

Mammoths

Giants of the Ice Age

ERROL FULLER

BUNKER HILL PUBLISHING
BOSTON

The Mammoth, *painted by the celebrated Czech palaeontological artist Zdenek Burian.*

Introduction

Mammoth, like dodo or shark, is one of those words that has come to mean something more than the animal it originally described. Shark, in addition to being the name of certain carnivorous fish, signifies rapacious; dodo not only identifies a species of extinct, flightless bird, it is also virtually a synonym for dead or useless; and mammoth means immense, overpoweringly large—there are "mammoth" sales, "mammoth" trucks, "mammoth" portions of cotton candy. The animal itself has passed into popular culture in a way that comparatively few creatures do. There are soft, cuddly toy mammoths, hard plastic ones, model mammoths in kit form, and mammoth films and cartoons. Like the dinosaurs or the dodo, this is one of the superstars of extinction, and it is probably true to say that although no one has ever seen a mammoth alive, it is more instantly recognized than the vast majority of living creatures.

The image of a shaggy, reddish, elephant-like behemoth trudging across wastes of ice and snow is familiar from innumerable paintings made during the last hundred years or so. These dramatic reconstructions are, of course, essentially works of the imagination, even though they may be conceived and based on a sensible interpretation of the available evidence. One of the curious things about the mammoth, however, is the fact that truly authentic pictures do exist. Images of mammoths are among the most ancient surviving products from the hand of man. Such antique images occur in surprisingly large numbers on cave walls across Europe, and it is clear that mammoths figured large in the minds of our prehistoric ancestors.

But why should an extinct elephant have acquired such celebrity today when other equally fascinating extinct creatures remain relatively obscure? Who knows, for instance, of the gigantic *Indricotherium*—a relative of the rhinoceros that once roamed the steppes of Asia, twice as large as an elephant—or the terrifying *Megalania*, an awesome twenty-foot-long lizard that may have survived in Australia until quite recently? Only enthusiasts. These are not creatures that have fired the public imagination, even though they are more than spectacular enough to have done so.

A toy stuffed mammoth.

Perhaps the key to the mammoth's popular success is the fact that it interacted so closely with our early ancestors. Perhaps it is its elephantine affinities—elephants being so familiar and well loved. But in addition to these similarities, the mammoth exhibits several intriguing extra features. First, there are the huge curving tusks. These could sometimes grow to truly remarkable lengths and weights, although the beast itself, contrary to popular supposition, was not significantly different in size to the modern-day African elephant. Then there is the fact that the mammoth was hairy and often reddish in color, whereas today's elephants are, essentially, bald (they actually have a light covering of hair) and greyish.

There are also certain contradictions connected with the mammoth that add to its aura of fascination.

The first of these is that the mammoth was an elephant that could survive—and even thrive—in cold climates. Today's elephants are, of course, creatures of the tropics. Just as there is something endearing about the idea of a furry elephant, so there is something entirely incongruous about the idea of an elephant in snow.

The probable date of the mammoth's extinction is also intriguingly contradictory. The animal may well have survived until just four or five thousand years ago—just as man was beginning to reach a level of true civilization. In other words, these creatures died out at a period in time that is almost close

A toy plastic mammoth.

The Indricotherium, *painted by Zdenek Burian.*

A beautiful fossil tusk.

A cartoon from the cover of the magazine Punch *showing a mammoth preserved in ice.*

enough for us to touch, yet remote enough to remain mysterious, and in any real sense, beyond reach. We have pictures contemporary with the mammoth's existence, but these have an alien quality and exist not in galleries or exhibitions, but in dark, eerie, inaccessible cave systems. They merely hint at a strange connection between man and beast. Finally, we have actual remains. Unlike most prehistoric creatures, the mammoth is known from more than fossilized bones. Entire bodies survive frozen and preserved in the wastes of Siberia, and from time to time one of these is found and made available for study.

What is a Mammoth?

In the long history of the world, there have been many different elephant-like species. Of these, only the elephants of Africa (*Loxodonta africana*) and India (*Elephas maximas*) now remain. All the rest are extinct, and among these extinct kinds are the mammoths.

Broadly speaking, mammoths can be defined as a small group of species sharing certain defining characteristics. These features include: a great domed forehead, small ears, huge curving tusks that have a beautiful twist, a tendency to furriness, a distinct patterning to the teeth, and a marked downward slope to the back. Although the mammoths have died out, leaving no descendants, it is clear that today's surviving elephants are fairly close relatives.

Reconstruction of a Mastodon. Drawing by Zdenek Burian.

Zoo elephants—the mammoth's surviving relatives. An African and Indian elephant photographed together. The much larger ears of the African can easily be seen.

There is often some confusion between "mammoth" and "mastodon," and it is sometimes assumed that these names are interchangeable. They are not. Mastodons are the product of an entirely different elephantine line that culminated in the species known as the American Mastodon (*Mammut americanus*). Like the last of the mammoths, this species became extinct only a few thousand years ago.

Zoologically speaking, "mammoth" is a somewhat loose term, there being several related species that can all be fairly described as "mammoths." The first and most ancient of these is known scientifically as *Mammuthus subplanifrons*, a creature identified from fossil remains found in Africa. *Mammuthus meridionalis* is a later ancestral mammoth from which the more celebrated species seem to be descended. *Mammuthus meridionalis* already showed the domed head and spiralling twist to the tusks that is a mammoth characteristic. It appears to have been a descendant of a more tropical species that probably migrated north from Africa to southern Europe.

The Columbian Mammoth Mammuthus columbi, *painted by Zdenek Burian.*

From that animal the Steppe Mammoth (*Mammuthus trogontherii*) evolved. This was the largest of all the species, standing some 14 feet [4.3 meters] at the shoulder. Its weight was probably around 10 tons.

It seems that either this species, or the ancestral mammoth *Mammuthus meridionalis*, crossed from northern Asia into North America by means of the land bridge that once connected the two continents. From this stock, the Columbian Mammoth (*Mammuthus columbi*) developed. This enormous creature—virtually as large as the Steppe Mammoth—occupied much of the North American continent, even spreading south into what is now Mexico. There is some argument over whether it diversified from one species into several. Whatever the truth may be, several scientific names have been suggested (*Mammuthus imperator* and *Mammuthus jeffersoni* among them) for the various supposed species.

In addition to these above-mentioned species, several dwarf mammoths have been described from islands as far apart as Malta in the Mediterranean and Wrangel Island in the Bering Strait.

When people think of "the" mammoth, however, it is the Woolly Mammoth (*Mammuthus primigenius*) that is usually under consideration. This is certainly the mammoth of popular imagination and is the species that is largely featured in this book. This descendant of the Steppe Mammoth seems to have evolved in Eurasia, but after spreading across the entire northern part of this vast area, it launched a second wave of expansion into North America across the same land bridge that its relatives had previously crossed.

The Steppe Mammoth Mammuthus trogontherii, *painted by Zdenek Burian.*

The Mammoth in Life

In its heyday—probably around 50,000 years ago—the Woolly Mammoth occupied a vast area. Its range stretched from the west coast of Ireland, across Eurasia and North America, until finally the species reached the Atlantic seaboard. There was no real barrier to such expansion. Britain was linked to continental Europe by a land bridge (many mammoth teeth are, in fact, dredged up by fishing vessels from the bed of what is now the North Sea), and Asia was similarly joined to North America. The English Channel and the Bering Straits were no obstacle to the movement of terrestrial mammals.

Within its range, the species seems to have inhabited the land just to the south of the great ice sheets, a location blessed with lush grassy vegetation. There is a general consensus of opinion—based around the development of the teeth and the kind of locality in which mammoth fossils are found—that the Woolly Mammoth was a grass eater. This is unusual for an elephant. Most eat leaves, fruits, bushes, and saplings. They even strip bark from trees. Given their necessarily enormous appetites, they are highly destructive to vegetation. In the colder climates inhabited by mammoths, such foodstuffs would have been limited, although clear northern skies, providing longer hours of sunlight combined with plenty of rain, would have promoted a rich growth of greenery. Grass may well have formed the staple diet, but we know that mammoths also fed on conifers, willows, birches, and alders, for the remains of these have been found in the stomachs of frozen specimens. How the beasts fed in winter is anybody's guess. Presumably, they moved southward in order to avoid the harshest weather, then traveled north again to take advantage of spring and summer growth. Comparison with modern day elephants, along with the evidence provided at fossil sites, seems to indicate that the animals lived in herds—perhaps very large ones. Given their enormous bulk (and, therefore, the constant need for large quantities of food), they must have gathered together at certain times to migrate in search of new feeding grounds.

The enormous hump on the mammoth's upper back, so visible in cave paintings, may have provided a fatty reserve that could be used up in times of shortage. When pressed, mammoths may also have been able to use their huge tusks to scrape away ice in search of vegetation.

A reconstruction inspired by the Beresovka Mammoth.

Mammoth—Vital Statistics

Woolly Mammoth
(*Mammuthus primigenius*)

Size

Height: around 10 feet at the shoulders [3 meters]
Length: $11\frac{1}{2}$–13 feet [$3\frac{1}{2}$–4 meters]
Trunk: $6\frac{1}{2}$ feet [2 meters]
Tusks: 10 feet [3 meters]

Weight 5-6 tons

Appearance

Like an elephant but with a distinctly sloping back and dense reddish hair. This hair could be up to $3\frac{1}{4}$ feet [1 meter] long and was coarse and thick. Underneath it was finer. The forehead showed high single-dome, the ears were noticeably small, and the tusks gigantic and twisted. The tail was short and the trunk nozzle had one short, "finger" and one long one to enable gripping movements. The skin was thick, about $\frac{3}{4}$ inch [2 cm] and covered a layer of fat that was up to 4 inches [10 cm] thick.

Extinction

In general terms this probably occurred around 8,000 years ago, although a dwarf kind of mammoth was still surviving on Wrangel Island (Bering Straits) around 4,000 years ago.

Other Species

Mammuthus meridionalis—
Ancestral Mammoth

Mammuthus columbi—
Columbian or American Mammoth

Mammuthus trogontherii—
Steppe Mammoth

Other names which sometimes occur in books and may or may not describe valid and distinct species

Mammuthus imperator

Mammuthus jeffersoni

Mammuthus subplanifrons

Mammuthus africanavus

Mammuthus exilis

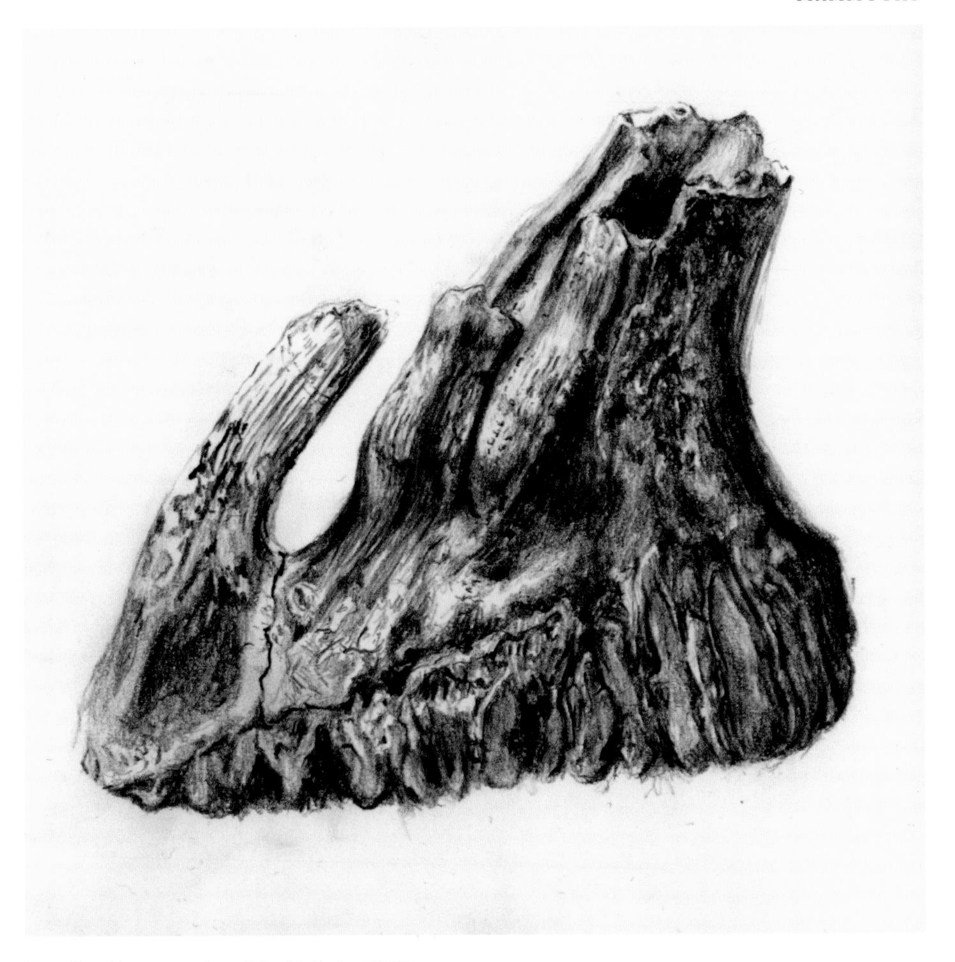

Drawing of a mammoth tooth by Catherine Wallis.

The Mammoth and the Ice Age

"Ice Age" is a commonly—but rather glibly—used expression. It conjures up visions of a remote past, when great snows and endless sheets of ice bleakly covered the land—a land that is now full of streams, meadows, and woodlands, and much more temperate in climate. We think of large mammals trudging despondently across these great desolate white wastes, or of our ancestors, dressed in skins and furs, huddling around fires that barely hold back the cold and gloom.

Yet the fact is that what we now call the Ice Age was by no means one-faceted. It was a highly complex event, or series of events,

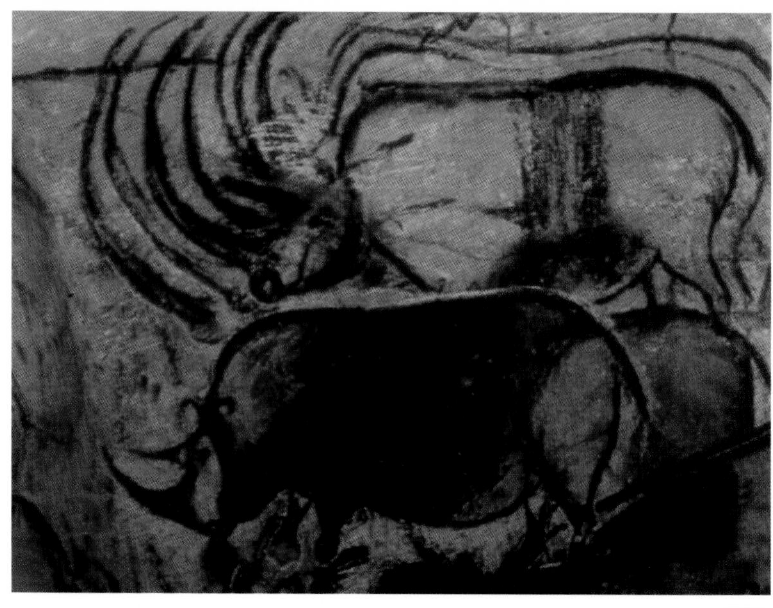

Cave painting of a Woolly Rhinoceros, a contemporary of the mammoth.

An Ice Age view. The Mammoth, *as imagined by the celebrated American palaeontological artist Charles R. Knight.*

in which great ice sheets and glaciers advanced from the north and then retreated—over and over again. Until recently, it was believed that there were perhaps four or five of these advances and retreats; now it is thought that there were many more. There would be warm periods between each advance, but these seem to have ended with surprising suddenness, and the ice would move southward again.

Naturally, a fauna evolved that could cope with the conditions imposed by climatic changes. The woolly rhinoceros, bison, reindeer, cave bears, and, of course, the mammoth were all part of this fauna. Such creatures could be found ranging over vast areas.

Most of the more spectacular of these animals are now gone, vanished with the cold climate that caused their emergence. Yet the last Ice Age ended only some 10,000 years ago, and there is no guarantee—despite dire warnings concerning the threat of global warming—that another is not on the way.

Frozen Mammoths

One of the more extraordinary aspects of mammoth history is the fact that, unlike most other prehistoric animals (which are known only from fossilized bones), entire mammoth carcasses are occasionally found, enabling us to form a very exact picture of how these great beasts actually appeared in life.

There are essentially two interlocking reasons for this unusual state of specimen preservation.

The first is connected with time. In comparative terms, the mammoth has not been extinct for very long. Dinosaurs, for instance, vanished from the earth around 70 million years ago, whereas mammoths have only been gone for a few thousand years. Indeed, 30,000 years ago they were probably common across vast parts of the northern world. Yet 30,000 years—or even 30—is in the ordinary course of events more than long enough for an elephantine body to rot away, leaving behind absolutely no trace of itself.

The other factor is a climatic one. Mammoths lived in conditions and situations that occasionally allowed their bodies to be speedily frozen after death. Because such events happened comparatively recently, at least in geological terms, some of the areas in which they occurred have not been subjected to climatic change, and the frozen carcasses have remained in deep freeze ever since. In many areas of permafrost, the ground has been frozen to a depth of 1,640 feet [500 meters] for thousands of years. During the brief Siberian summer, the thaw affects only the top 6½ feet [2 meters].

Every so often, local changes in the environment or slight alterations in temperature mean a thaw—and a carcass, frozen for centuries, begins to emerge from the Siberian permafrost. The meat can be so fresh that local people have been known to feed it to their dogs. In such harsh and unforgiving terrain, where no potential food source can be wasted, such people have undoubtedly sometimes sampled the ancient flesh personally! Given the sparse population in these remote, inhospitable areas, many frozen mammoths must have emerged from the frost only to be eaten by scavengers and insects before any human being could see them.

There are probably thousands of mammoth bodies, in various states of completeness, still lying undisturbed in the wastes of Siberia; certainly, there has been a considerable trade in mammoth ivory, found and

THE Sun

Thursday, November 4, 1999 30p THOUGHT: GINGER TWICE

PICTURE EXCLUSIVE

THE MAMMOTH

How it was prised from ice after 20,000 years

SEE PAGES 32 & 33

cture: LATREILLE/J FRASER

The front page of The Sun *(November 4, 1999) featuring the discovery of a frozen mammoth.*

The frozen mammoth, still in its icy tomb, being lifted by helicopter.

then marketed by the local people for centuries, and an awareness of the potential for mammoth discoveries has long existed among Russian naturalists. As long ago as 1860, the St. Petersburg Academy of Sciences posted advertisements requesting information that might lead to the acquisition of specimens. One hundred roubles was promised for help in the recovery of complete skeletons; considerably more was offered for an entire carcass.

Very few entire mammoths are reported, however, and even fewer are actually collected for museums or other scientific institutions. The reasons for this are complex and curious.

Local people quickly discovered that, despite financial inducements, the disadvantages of having palaeontologists and other scientists descend on their communities outweighed any advantages. General interference added to imperious demands on local manpower, and transport resources were insufferably limited. There were also certain superstitious beliefs and taboos that inhibited the locals from handling dead mammoths.

In addition to such limitations, there are more obvious practical problems. The difficulties involved in removing, from a desperately remote place (often in a particularly harsh environment), a partially frozen crea-

ССРС НАУКАЛАРЫН АКАДЕМИЯТА
СИБИРДЭЭҔИ САЛАА САХА СИРИНЭЭҔИ ФИЛИАЛА
ЯКУТСКАЙ К.

СИР АННЫТТАН ХОСТОНОР КЫЫЛЛАР ТУСТАРЫНАН

Саха АССР өрүстэрин хочолоругар уонна Муустаах океан кытылыгар мамоннар, носорогтар, дьиикэй оҕустар, сылгылар уонна былыр Сибиргэ үөскүү сылдьыбыт атын да кыыллар өлүктэрэ сир анныттан ирэн тахсаллар.

Хас биирдии оннук өлүк эттэри-тириилэри, түүлэри булуллубута наукаҕа улахан суолталаах.

Саха сирин булчуттара, балыксыттара, сири хаһар тэриллэр үлэһиттэрэ уонна да атын олохтоохторо оннук булумньулар үөрэтэргэ-чинчийэргэ уонна музейдарга туруорарга олус наадалаахтарын сороҕор билбэттэр.

Ол иһин СССРС Наукаларын Академиятын Сибирдээҕи салаатын Саха сиринээҕи филиала сир анныттан кыыл өлүгэ ирэн таҕыстыбытын булбут эбэтэр истибит гражданнартан бардарыттан көрдөһөр: бииринэн, ол өлүктэри сытыйыыттан уонна сиэмэх кыыллартан харыстыыр туһугар дьаһал ыларга, иккиһинэн, оннук булумньу туһунан 677007 г.Якутск, Якутский филиал СО АН СССР диэн аадырыска телеграбынан эбэтэр почтанан биллэрэ охсорго.

Биллэрбит дьону ССРС НА СС Саха сиринээҕи филиала Бочуоттуһай грамоталарынан наҕараадалыаҕа уонна булумньу наукаҕа төһө суолталааҕынан көрөн харчынан манньалыаҕа.

ССРС Наукаларын академиятын Сибирдээҕи
салаатын Саха сиринээҕи филиалын Президиумун председателэ ССРС НА член-корреспондента
Н.В.Черскэй

АКАДЕМИЯ НАУК СССР
ЯКУТСКИЙ ФИЛИАЛ СИБИРСКОГО ОТДЕЛЕНИЯ, Г. ЯКУТСК

О НАХОДКАХ ИСКОПАЕМЫХ ЖИВОТНЫХ

В долинах рек и на побережье Ледовитого океана на территории Якутской АССР вытаивают из земли трупы мамонтов, носорогов, диких быков, лошадей и других животных, когда-то живших в Сибири.

Каждая находка такого трупа с мясом, кожей и шерстью представляет большой научный интерес.

Охотники, рыбаки, работники горных предприятий и другие жители Якутской республики не всегда знают, что такие находки очень нужны для их научного изучения и выставки в музеях.

Поэтому Якутский филиал Сибирекого отделения Академии наук СССР просит всех граждан, нашедших вытаивающие из земли трупы животных или слышавших о них, во-первых, принять меры к их охране от дальнейшего вытаивания и уничтожения хищными животными, а во-вторых, возможно скорее сообщить об их находках телеграфом или почтой по адресу: 677007 г. Якутск, Якутский филиал СО АН СССР.

За это Якутский филиал СО АН СССР будет награждать заявителей Почетными грамотами и выдавать денежные вознаграждения, в размерах, зависящих от научной ценности находки.

Председатель Президиума
Якутского филиала Сибирского
отделения Академии наук СССР
член-корреспондент АН СССР
Н.В.Черский

Просим повесить это обращение на видных местах в ваших поселках.

A notice distributed by the old Soviet government soliciting mammoth remains. The Yakutian language is used on the left and Russian on the right.

ture that might weigh in the neighborhood of 10 tons are easy to imagine. In many cases, the job has to be done with great speed, and by the time word of a discovery reached the appropriate palaeontologists, the find might well be ruined due to decomposition or a general dismantling by scavengers. Then there is the matter of money. The financial resources needed to excavate, transport, and preserve such specimens are enormous.

The most commonly asked question about these preserved mammoths is "How did they die?" There seems to be no simple answer. There is little evidence that these

A female baby mammoth that was named Mascha.

Hosing out the Shandrin Mammoth, found in 1972. The water jet was used to melt the permafrost.

The Shandrin Mammoth after hosing

creatures actually froze to death, for the bodies appear to have been in reasonably good shape, with full stomachs. Some even show the remains of food in their mouths. Probably, they died in all sorts of ways: some by drowning, or being overwhelmed in a mudflow or swamp, some by crashing through thin ice, or falling into crevasses or over cliffs. The factor linking all together is the rapidity with which the cold descended. Presumably, all died in late autumn, and their bodies were covered by ice, or by mud that then quickly froze, leaving them refrigerated for millennia.

The Adams Mammoth

The first frozen mammoth discovery for which we have a definite and detailed record is a carcass that over the years has come to be known as "The Adams Mammoth." The story of its discovery and collection was spread over almost a decade, but it began during 1799, when a Siberian ivory dealer known as Ossip Shumakhov spotted several gigantic blocks of icy material by the edge of a swamp at the mouth of the River Lena. One of these blocks had a particularly strange and amorphous quality to it. Despite dire warnings from his wife and family—who had heard of such things before and feared they were portents of disaster—Ossip kept the vast mass under periodic observation, and by 1801 he noticed that one whole side of a mammoth and one of its tusks were plainly visible. Two years later, a large part of the carcass fell away onto a sandbank, and Shumakhov could resist no longer. He sheared away the tusks and sold them.

Three more years were to pass before there was any further development. Then, during 1806, a botanist from the St. Petersburg Academy of Sciences by the name of Mikhail Adams visited the site and found that the mammoth was still there. Much of the flesh had gone—eaten by dogs, wolves

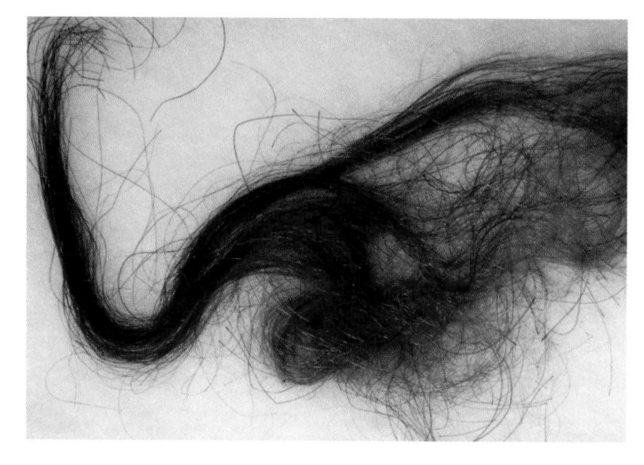

Hair from a frozen mammoth.

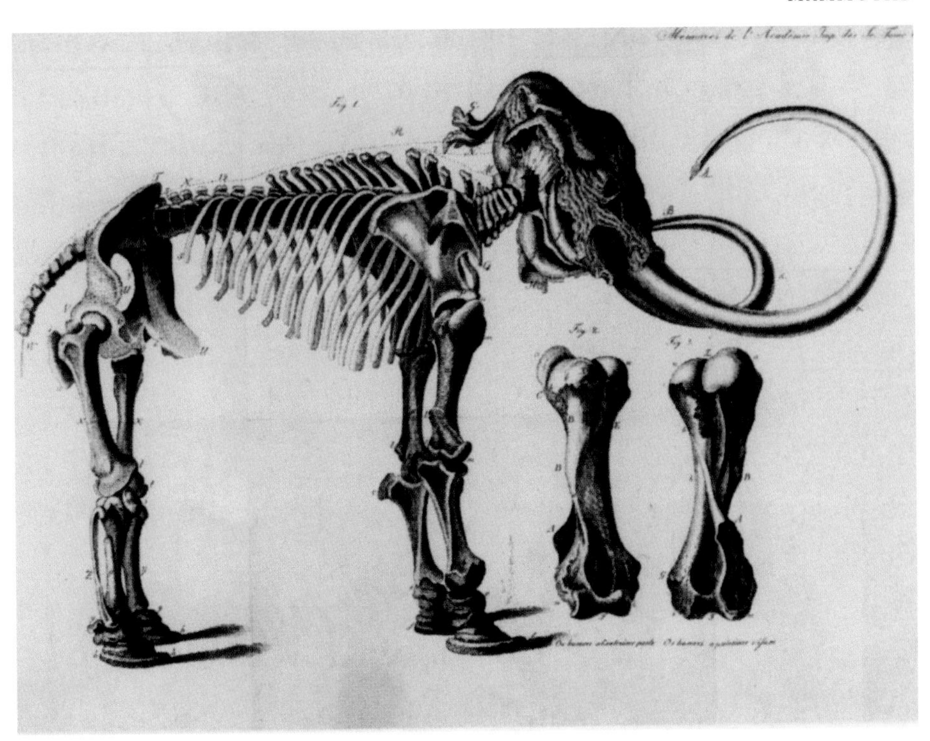

Lithograph showing the skeleton of the Adams' Mammoth.

and foxes, bears and wolverines—but Adams collected what was left, along with the entire skeleton. He also managed to locate the new owner of the tusks, which were over 10 feet [3 meters] long, and purchased them back.

By 1808 he'd achieved the remarkable feat of transporting the material back to St. Petersburg and reassembling it for display. It measured some 16½ feet [5 meters] in length and was 10 feet [3 meters] in height. Scraps of skin and strands of hair were circulated to museums all around the world

The Beresovka Mammoth

The most celebrated of all mammoth carcasses is known as the Beresovka Mammoth. Its story is a particularly curious one.

During the year 1900, a Cossack trader reported to the then Governor of Yakutsk—a Russian province in Siberia—that he had purchased two great tusks from a Lamut tribesman. The report didn't end there, however, for with the tusks came a story. They had been cut from an enormous beast frozen in the ice. The Lamut had dared this much, but feared to disturb the monster further—

The Beresovka Mammoth as it looked after it had been restored and prepared for exhibition.

Two views of the Bereskova Mammoth as it appeared when the recovery expedition reached it.

either from superstitious traditions or out of a genuine and understandable fear of the unknown. But the tribesman had willingly disclosed to the Cossack trader the exact location where he'd made his find, and the Cossack in turn happily passed this information on to the provincial governor. He, also in turn, felt obliged to notify the authorities in far-off St. Petersburg.

His written report indicated that a gigantic carcass was frozen into the side of a cliff by a river known as the Beresovka. This locality is well inside the Arctic Circle, and it appears that erosion or perhaps slight changes in the course of the river, had caused something of a landslide, and this had revealed the creature's presence.

An expedition was speedily organized to hunt down this exciting find, and on May 3, 1901, a team led by two zoologists—Otto Herz and Eugene Pfizenmayer—left St. Petersburg by train, heading first for Irkutsk, then continuing on east and north. It took the team until September 9 to even reach the site—more than four months of travel!

During the time that had elapsed since the initial discovery, much damage had, understandably, been done to the remains. Most of the flesh on the exposed head had been eaten to the bone by scavengers, and some of the internal organs had rotted away. A great deal of the carcass, however, remained in good condition.

With winter coming on, the investigators had to work fast. They spent six weeks removing great quantities of flesh from the creature's hindquarters and head, and these dainties they wrapped in hides and re-froze. Meanwhile, they built a great hut over the remains. Pfizenmayer recalled that the vicinity of the mammoth smelled like a "badly kept stable blended with the stench of offal."

By October 11, the party had completed their work and were ready to leave with the booty. Had they stayed any longer, their return journey would have probably been impossible until the following spring. The temperatures had already plummeted to well below zero, and in any event, they didn't get back to St. Petersburg until February, 1902.

The specimen was recreated—using as

The hut that was built over the remains of the Beresovka Mammoth.

much original material as possible—and put on show, becoming one of the great treasures of the St. Petersburg Zoological Museum. Immediately, it aroused a great deal of interest. The ill-fated Tsar and Tsarina, Nicholas II and his wife Alexandra, were among the first to visit, the Tsar himself listening intently to accounts of the finding, while Alexandra pressed a handkerchief delicately to her nose.

It is estimated that the beast itself died some 30,000 years ago. Researchers believe that it probably fell into a crevasse (several bones were broken) from which it was unable to extricate itself. Some physical evidence indicates that it died of asphyxiation, and that death was relatively swift; remains of food were found in the animal's mouth.

A local Yakut holding up part of the mammoth's skin.

Part of a leg of the Beresovka Mammoth showing muscles and tendons.

Presumably, the body was frozen quickly due to the rapid onset of winter, and this fact alone was probably responsible for the mammoth's good state of preservation when it was first found. This particular animal has been estimated as being just under 40 years of age at the time it met its death.

In 1908, Eugene Pfizenmayer returned to the area northeast of Yakutsk to recover the remains of another mammoth. His reappearance—in perfectly good health—astonished the locals, who believed that his part in interfering with the dead mammoth would surely have occasioned his death. They were, however, bolstered in their superstition by the discovery that Otto Herz was not with the new party. He had, in fact, died during the intervening years.

 Dima the baby mammoth in situ.

Dima the baby mammoth

A number of more or less complete baby mammoths have been found, and the most famous of these is "Dima." Dima's discovery generated a great deal of publicity all around the world, and indeed, the preserved specimen was sent on a world tour and exhibited in museums in Japan, North America, and Europe. During the tour, this baby mammoth was given an insurance value of $12,000,000 [around £8,000,000].

He was found in June of 1977 by a bulldozer driver named Alexei Logachev, who noticed the carcass while bulldozing through a mass of frozen silt by the side of a small tributary of the River Berelekh. Here, in a place known as the Kirgilyakh Creek, Dima—named after a local stream—lay at a depth of about 6½ feet [2 meters], so hidden from view that the plough of the bulldozer cut away part of its side before the driver spotted it.

Logachev, who was engaged in a gold-mining project, received 1,000 roubles for his part in the discovery, plus a silver medal from the St. Petersburg Academy of Sciences. His fellow workers were also compensated for the inconvenience of suspending their operations while the find was properly investigated and removed.

It is estimated that this creature was no more than a year old, and perhaps only six months, when it died—around 40,000 years ago. It would have stood about a meter [3¼ feet] in height, its length being just a little greater. At the time of death, Dima seems to have been emaciated, with various physical features indicating that he was in poor general condition. The little creature may have become stuck in a swamp or mudflow, or it may have fallen into a crevasse from which there was no escape. Alternatively, it may have become separated from its herd and wandered aimlessly until overwhelmed by exhaustion, starvation, or cold. The stomach contained no plant remains, only silt and the baby's own hair, which seems to indicate a certain desperation of circumstance.

Whatever actually happened, Dima was found in an almost perfect state of preservation, indicating that he was covered with either silt or ice immediately after death and rapidly frozen. As the body remained covered until the moment of its discovery, it was not subjected to the usual attacks by scavengers. Dima is probably the most perfectly preserved of all mammoth carcasses—even his blood cells remain intact.

Another view of Dima as he was found.

Dima as he looked after being preserved.

Dima in London during his world tour.

Other Mammoth Remains

As might be expected, frozen mammoths constitute only a very small proportion of the mammoth remains that exist in the world's museums and in private collections. Bones and other relics are found across much of North America, Europe, and northern Asia, in all sorts of localities and situations. In fact, mammoth bones, parts of tusks, and teeth are comparatively common fossils. Mammoth ivory is by no means rare, and large quantities of bones have been found in certain locations. In some places, the remains are so plentiful that shelters and huts have been constructed from them by local peoples who found building materials in short supply. Teeth—and sometimes skeletal elements—are regularly hauled up from areas that were once firm land by fishing vessels dredging in shallower parts of the North Sea. Such teeth, in particular, are fairly common and are often seen for sale in fossil or curio shops.

Indeed, the teeth are of considerable importance in determining the progress of mammoth evolution. Ancestral mammoths

A native Yakut with the lower jaw of a mammoth

A particularly large tusk found in Siberia during the early decades of the twentieth century.

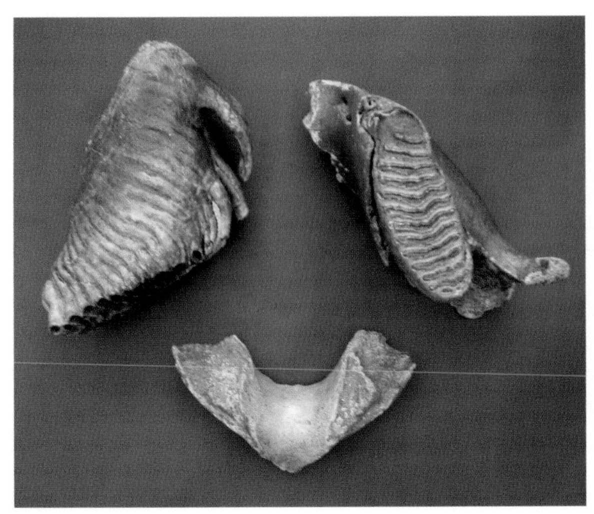

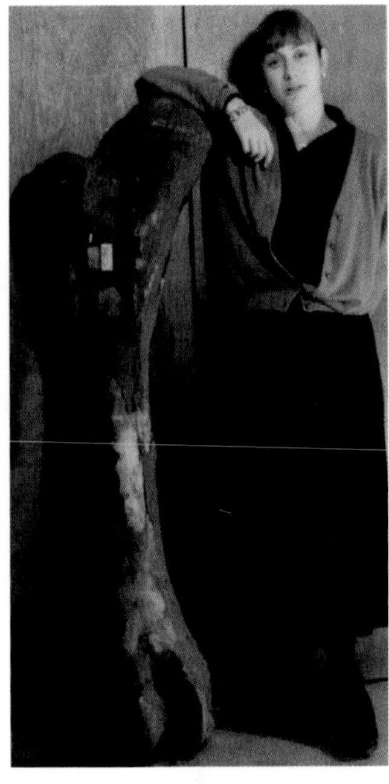

Left: Mammoth teeth are comparatively common fossils. This photograph shows several, along with the tip of a mammoth's jaw.

Right: The world's (allegedly) largest mammoth thigh bone, found during the nineteenth century at Mundesley, Norfolk, England.

had teeth with lower crowns and fewer enamel ridges (typically 12 to 14) along the top. Such tooth design provided moderate chewing power, indicative of a diet that probably consisted of leaves and the soft parts of shrubs. Later mammoths, and in particular the Woolly Mammoth, have higher crowns and many more enamel ridges (up to 26). This suggests a shift in emphasis to a more grassy diet, which is harsher on the teeth, causing quicker wear because of the extra chewing required and the tendency for grit and dirt to be scooped up along with the grass.

To offset the effect of the higher crowned teeth, it became necessary for both upper and lower jaws to deepen and the top of the skull to become enlarged. This last feature also helps to balance and support the enormous development of the tusks.

Stone Age Images

Portraits of mammoths—pictures that have survived for thousands of years—are among man's earliest artistic endeavours. Extraordinarily, these images occur, along with those of other animals, on the walls of many caves across southern Europe, caves that were either inhabited, or visited regularly, by our Stone Age ancestors. Sometimes, they are in alarmingly perfect states of preservation.

Often, they occur deep within cave systems, and not—as might be expected—at or near the caves' mouths. Why were they made, and when, and under what conditions? These are questions often asked; but the answers are, for the most part, speculation. We know only that the images exist, and that the best of them provide a fairly clear idea of what the creatures depicted actually looked like. Some show considerable sophistication in terms of technical ability, while others are less satisfactory.

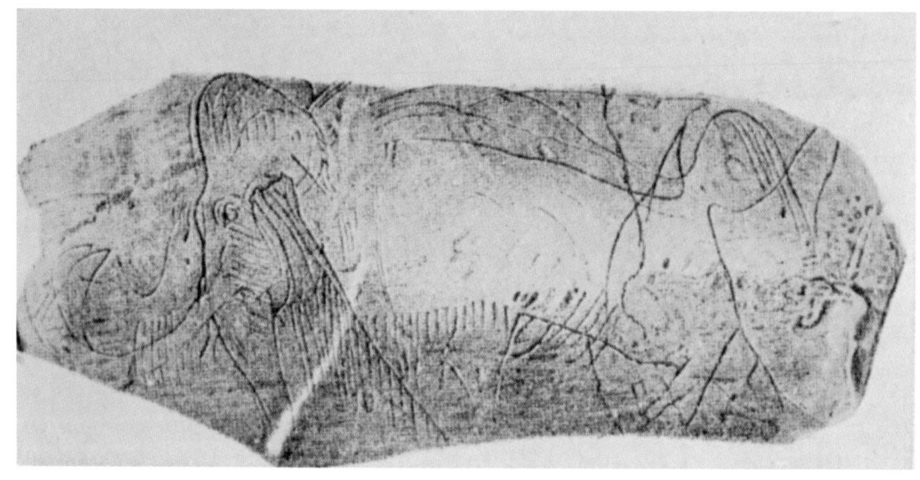

Prehistoric engraving on a piece of mammoth ivory found at La Madeleine, France.

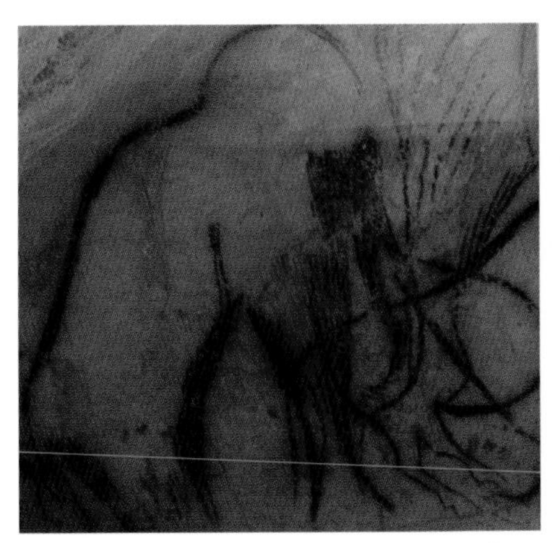

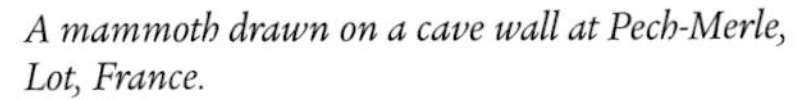

A mammoth drawn on a cave wall at Pech-Merle, Lot, France.

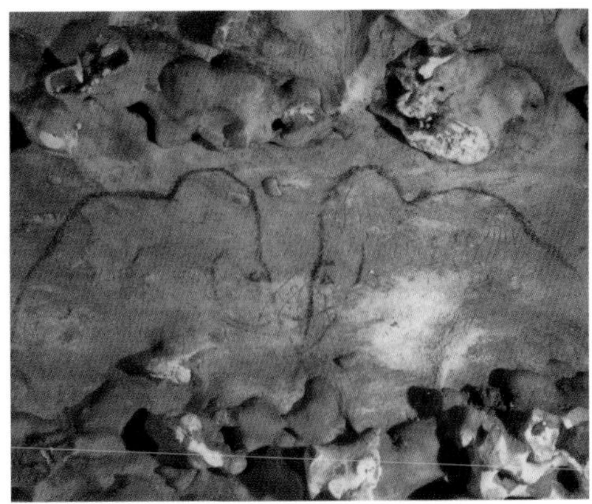

Two mammoths on a wall in Rouffignac Cave, France.

As far as mammoth images are concerned, these often are seen alongside pictures of other prehistoric creatures—bison, horses, reindeer, etc.—and the animals are almost always drawn from the side.

Although we can never be truly sure of their purpose, or of what part they played in the lives of those who made them, they seem clearly to have been creations of considerable importance. Why else would they be present in so many places? There are, of course, many theories about the intentions of both those who painted the pictures and those who looked at them. The most likely idea is that they served some function in early man's hunting plans.

The idea of possession plays a large part in the concept of art—the painting or drawing of a place, a person, or a creature being, to a greater or lesser degree, an act of artistic possession. Similarly, to convincingly reduce a three-dimensional object to a two-dimensional one might be considered an act of magic. Put these two concepts together and you have a recipe for organized ritual. There is a modern assumption that the prehistoric act of painting creatures in a convincingly lifelike manner—or even in a stylized one—conferred upon the painter, and perhaps his audience as well, a control over the object of artistic interest. In more practical terms, perhaps the artist was ensuring success at the

Engraving at Rouffignac Cave, France.

A bas-relief from the "Mammoth Cave" at Domme, France.

hunt, or protection from those creatures that might be considered dangerous. In either case, this was a matter of life and death.

Whether or not these theories explain the purpose behind the images, the pictures remain to show us exactly what the mammoth looked like in life. They provide confirmation of what we know from the remains of mammoths preserved in ice.

It is unknown whether prehistoric man also drew mammoths and other animals in places besides caves—on rock faces, bark, or skins, for instance. Such things have not survived the test of time. But perhaps they did, for we do have images scratched on, or sculpted from, pieces of bone or stone. In a few cases, mammoths are the subject.

Dating mammoth pictures is as controversial as determining their purpose. They seem to have been painted over a period of thousands of years, but actually dating them, even using modern scientific techniques, is not easy. The best guesses indicate that some may be 30,000 years old, while many were probably produced between 12,000 and 15,000 ago.

A highly stylised mammoth at La Baume Latrone, France.

The prehistoric burial of a chief, imagined by the Czech painter Zdenek Burian. A mammoth shoulder bone is about to be put on the corpse, and a tusk lies at the side of the grave.

A hut made from mammoth bones. The hut—one of several found at Mezhiridi, Ukraine—has collapsed at some time during its 15,000 years' existence.

The hut after museum reconstruction.

Extinction

Modern mammoth "hunters" bearing an ivory trophy.

Thirty-thousand years ago, great populations of mammoths roamed across vast areas of Europe, Asia, and North America, just as they had done for tens of thousands of years. Then, with a terrible swiftness (at least in terms of geological time), the great herds dwindled to localized bands of animals that grew ever smaller and less significant. Within another 20,000 years, the mammoth was virtually gone. Radiocarbon dating on fossils, backed up by the scarcity of remains, indicates that 10,000 years ago the mammoth was a rare creature, and at an unknown date in the ensuing millennia it became extinct.

A few dwarf mammoths survived for a while longer on Wrangel Island off the coast of Alaska, and there may possibly have been pockets of survival elsewhere, but to all intents and purposes the mammoth was gone.

Why? We don't really know, although it is possible to make two very good guesses, and one of them almost certainly provides the answer.

The influence, and hand, of man is probably the single most compelling cause of the mammoth's extinction. Just as the habits of the modern day African elephant make it incompatible with the requirements of man, so, probably, did the ways of mammoths conflict with the needs of our ancestors. Elephants take up too much space, cause too much damage to the environment, and in general terms, provide too big and tempting a target. In the last century or so, the numbers of African elephants have fallen drastically, and although there may still be many elephants in existence, the current population is a mere remnant of what it once was.

So, too, it probably went with the mammoth. The species used territory required by man. Not only did he hunt the mammoth—perhaps preying on the younger individuals that presented an easier target—but he also had a major impact on the environment, rapidly utilizing and altering the kind of territory that mammoths needed in order to thrive.

The other often-suggested cause for

Mammoth and Woolly Rhino—two icons of extinction. Watercolour by Maurice Wilson.

extinction is climatic change. During the period of mammoth decline, the great ice sheets were in retreat, and the climate became wetter and warmer. Clearly, this would have had an effect on the kind of vegetation (and its extent) that mammoths required. Even though much suitable habitat remained, it may have become fragmented, causing creatures as large as mammoths to be unable to sustain viable populations. This idea makes a good deal of natural sense, but there is one snag. The ice had retreated and advanced many times during the reign of the mammoth, and the species had previously been able to cope with the consequences.

What factor, then, made the difference in this final instance? Inevitably, we come back to the hand of man. The mammoth was by no means the only large mammal species to become extinct during this period. There were many others—the Irish elk, the cave bear, the woolly rhino, and the giant sloth are just a few. It cannot be a coincidence that the disappearance of all these great beasts coincided with the rise of man and the beginnings of his technological advance.

A Living Mammoth?

During 1987, rumors emerged from a remote area of Nepal that an elephant of truly enormous proportions, and with a huge, domed head, had been seen. Was it a mammoth? The descriptions matched, and were unequivocal, but the zoological world generally greets such sightings—Loch Ness monsters, yetis, and sea serpents, etc.—with a degree of skepticism.

Some people were persuaded, however, and an expedition was organized to settle the matter one way or another. Usually, such expeditions fail resoundingly to achieve their purpose. The location is reached, nothing is seen, the native people are questioned to tantalizing but inconclusive effect, the intrepid explorers retreat home empty-handed, vowing to return again at the earliest opportunity (they almost never do!), and the matter remains unresolved.

In this instance, however, the expedition—fronted by John Blashford-Snell and actress Rula Lenska—finally located their quarry. After weeks of searching, they saw and photographed the enormous beast in its jungle stronghold.

Yes, the creature in question did exist. Yes, it was remarkable. Yes, it looked as described. But, no, it was not a mammoth. DNA analysis of dung and other material revealed that the massive creature—along with several others living in the area—shared the same genetic sequencing as ordinary Indian elephants. Perhaps as a result of

The dust jacket of Mammoth Hunt.

isolation from other populations, combined with the small number of individuals, the beasts were developing characteristics that superficially resembled some of the features exhibited in mammoths.

But this is not the only tale of supposed living mammoths. For centuries, stories have emanated from Siberia about a great burrowing beast as large as a house, which lives underground but dies immediately if it dares to emerge from its hole, exposing itself to the rays of the sun or the moon. Doubtless, these tales are a result of simple people trying to make sense of the occasional discovery of frozen mammoth carcasses that appear to be in the act of emerging from a subterranean realm.

There are, however, more inexplicable tales. In his influential book *On the Track of Unknown Animals* (1958), Bernard Heuvelmans recounts the following story in a chapter called "The Mammoth of the Taiga." (The taiga is a vast forested area of Siberia and northeastern Asia that is virtually unexplored; it is about half the size of the United States). The story came originally from a diplomat working at the French consulate in Vladivostock around 1920. He told the story of an illiterate Russian hunter who had returned to civilization after spending some four years in the taiga:

The dust jacket of On the Track of Unknown Animals.

> *There had been a few big snowstorms, followed by heavy rain. It wasn't freezing yet, the snow had melted and there were thick layers of mud in the clearings. . . . It was in one of these clearings that I was staggered to see a huge footprint pressed deep into the mud. . . . The track suddenly turned east and went into the forest of middle-sized elms. Where it went in I saw a huge heap of dung; I had a good look*

at it and saw it was made up of vegetable matter. . . . I followed the track for days and days. Sometimes I could see where the animal had stopped in some grassy clearing and then had gone on forever eastward. Then, one day, I saw another track, almost exactly the same. It came from the north and crossed the first one. It looked to me from the way they had trampled all over the place . . . as if they had been excited or upset at their meeting. Then the two animals set out marching eastwards. I followed them . . . afraid for indeed I didn't feel I was big enough to face such beasts alone. One afternoon it was clear enough . . . that the animals weren't far off. . . . All of a sudden I saw one of the animals quite clearly, and now I must admit I really was afraid. It had stopped among some young saplings. It was a huge elephant with big white tusks, very curved. It was a dark chestnut colour as far as I could see. It had fairly long hair on the hind quarters but it seemed shorter on the front.

Frightened, and with evening coming on, the hunter left and sought shelter, but not

The Mammoth, *painted by the American artist Charles R. Knight.*

Large Siberian tusks photographed during the 1920s.

> *before he'd caught glimpses of the second animal. Winter was on its way, and he could afford no more time (nor, now, did he have the inclination) to track the mammoths.*

Such a story must either be accepted or rejected as it stands, for there is no follow-up to it. In all probability this tale is just that—a tale!

So is the mammoth gone forever? Perhaps not.

Today, the world of science opens up new horizons. *Jurassic Park* may be just a story, but many scientists now believe that DNA technology may facilitate the re-creation of some extinct species. If so, what better candidate could there be than the mammoth? The creature was truly spectacular, and there is plenty of surviving tissue to provide a basis for such a work. Perhaps the mighty mammoth will one day rise again, trumpeting from the grave.

Further Reading

Augusta, J. and Burian, Z. 1960. *Prehistoric Man.* London: Paul Hamlyn.

Augusta, J. and Burian, Z. 1966. *The Age of Monsters.* London Paul Hamlyn.

Bahn, P. and Vertut, J. 1988. *Images of the Ice Age.* New York: Facts on File.

Blashford-Snell, J. and Lenska, R. 1996. *Mammoth Hunt.* London: HarperCollins.

Digby, B. 1926. *The Mammoth and Mammoth-Hunting in North East Siberia.* London: Witherby.

Heuvelmans, B. 1958. *On the Track of Unknown Animals.* London: Rupert Hart-Davis.

Lister, A and Bahn, P. 1995. *Mammoths.* London: Boxtree.

Silverberg, R. 1970. *Mammoths Mastodons and Man.* New York: World's Work.

Sutcliffe, A. 1983. *On the Track of Ice Age Mammals.* London: British Museum (Natural History).

Swinton, W. 1966. *Giants Past and Present.* London: Robert Hale.